DOT-VNTSC-FAA-10-16
DOT/FAA/AR-TN10/28

Human Factors Considerations for the Integration of Traffic Information and Alerts on an Airport Surface Map

Human Factors Research and
Engineering Group, AJP-61
Office of Research and
Technology Development
Washington, DC 20591

Michelle Yeh, Ph.D.
Scott Gabree, Ph.D.

U.S. Department of Transportation
Research and Innovative Technology Administration
John A. Volpe National Transportation Systems Center
Cambridge, MA 02142

Technical Note
December 2010

Notice

This document is disseminated under the sponsorship of the Department of Transportation in the interest of information exchange. The United States Government assumes no liability for its contents or use thereof.

Notice

The United States Government does not endorse products or manufacturers. Trade or manufacturers' names appear herein solely because they are considered essential to the objective of this report.

REPORT DOCUMENTATION PAGE

Form Approved
OMB No. 0704-0188

Public reporting burden for this collection of information is estimated to average 1 hour per response, including the time for reviewing instructions, searching existing data sources, gathering and maintaining the data needed, and completing and reviewing the collection of information. Send comments regarding this burden estimate or any other aspect of this collection of information, including suggestions for reducing this burden, to Washington Headquarters Services, Directorate for Information Operations and Reports, 1215 Jefferson Davis Highway, Suite 1204, Arlington, VA 22202-4302, and to the Office of Management and Budget, Paperwork Reduction Project (0704-0188), Washington, DC 20503.

1 AGENCY USE ONLY (Leave blank)	2 REPORT DATE December 2010	3 REPORT TYPE AND DATES COVERED Technical Note, December 2010

4 TITLE AND SUBTITLE Human Factors Considerations for the Integration of Traffic Information and Alerts on an Airport Surface Map	5 FUNDING NUMBERS FA07C2 JT084 FA07C2 HT084

6 AUTHOR(S) Michelle Yeh and Scott Gabree	

7 PERFORMING ORGANIZATION NAME(S) AND ADDRESS(ES) U.S. Department of Transportation John A. Volpe National Transportation Systems Center Research and Innovative Technology Administration Cambridge, MA 02142-1093	8 PERFORMING ORGANIZATION REPORT NUMBER DOT-VNTSC-FAA-10-16

9 SPONSORING/MONITORING AGENCY NAME(S) AND ADDRESS(ES) U.S. Department of Transportation Federal Aviation Administration Office of Research and Technology Development Human Factors Research and Engineering Group, AJP-61 800 Independence Avenue, SW Washington, D.C. 20591 Program Manager: Dr. Tom McCloy	10 SPONSORING/MONITORING AGENCY REPORT NUMBER DOT/FAA/AR-TN10/28

11 SUPPLEMENTARY NOTES

12a DISTRIBUTION/AVAILABILITY STATEMENT This document is available to the public through the National Technical Information Service, Springfield, Virginia 22161.	12b DISTRIBUTION CODE

13. ABSTRACT

The purpose of this document is to provide human factors considerations in the integration of traffic information and indications and alerts for runway status on an airport surface moving map. The US DOT Volpe Center, in support of the Federal Aviation Administration (FAA) Office of Aircraft Certification, gathered information in two ways. First, we conducted observations during demonstrations of ADS-B surface conflict detection algorithms sponsored by the FAA Surveillance & Broadcast Services Office. Second, we visited Embry-Riddle Aeronautical University (ERAU) in Daytona Beach, Florida, to understand the impact of a surface moving map with ADS-B in the general aviation operating environment.

The results of this effort provide a preliminary glimpse into potential human factors concerns with the use of a surface moving map, traffic function, and the presentation of surface indications and alerts. The findings are organized into seven categories: use of color, indications/alerts, symbols, information prioritization, airport database, air-ground integration, and other. For each category, a description of the human factors concern, examples, and recommendations for addressing the concern are provided. The information is intended to support the development of minimum operational performance standards for surface conflict detection and alerting. Additional research will be needed to empirically determine appropriate design solutions.

14 SUBJECT TERMS Airport surface moving map, human factors, usability, traffic display, Automated Dependent Surveillance – Broadcast (ADS-B), surface conflict detection	15 NUMBER OF PAGES 34
	16 PRICE CODE

17 SECURITY CLASSIFICATION OF REPORT Unclassified	18 SECURITY CLASSIFICATION OF THIS PAGE Unclassified	19 SECURITY CLASSIFICATION OF ABSTRACT Unclassified	20 LIMITATION OF ABSTRACT

NSN 7540-01-280-5500

Standard Form 298 (Rev. 2-89)
Prescribed by ANSI Std. 239-18
298-102

METRIC/ENGLISH CONVERSION FACTORS

ENGLISH TO METRIC

LENGTH (APPROXIMATE)
- 1 inch (in) = 2.5 centimeters (cm)
- 1 foot (ft) = 30 centimeters (cm)
- 1 yard (yd) = 0.9 meter (m)
- 1 mile (mi) = 1.6 kilometers (km)

AREA (APPROXIMATE)
- 1 square inch (sq in, in^2) = 6.5 square centimeters (cm^2)
- 1 square foot (sq ft, ft^2) = 0.09 square meter (m^2)
- 1 square yard (sq yd, yd^2) = 0.8 square meter (m^2)
- 1 square mile (sq mi, mi^2) = 2.6 square kilometers (km^2)
- 1 acre = 0.4 hectare (he) = 4,000 square meters (m^2)

MASS - WEIGHT (APPROXIMATE)
- 1 ounce (oz) = 28 grams (gm)
- 1 pound (lb) = 0.45 kilogram (kg)
- 1 short ton = 2,000 pounds (lb) = 0.9 tonne (t)

VOLUME (APPROXIMATE)
- 1 teaspoon (tsp) = 5 milliliters (ml)
- 1 tablespoon (tbsp) = 15 milliliters (ml)
- 1 fluid ounce (fl oz) = 30 milliliters (ml)
- 1 cup (c) = 0.24 liter (l)
- 1 pint (pt) = 0.47 liter (l)
- 1 quart (qt) = 0.96 liter (l)
- 1 gallon (gal) = 3.8 liters (l)
- 1 cubic foot (cu ft, ft^3) = 0.03 cubic meter (m^3)
- 1 cubic yard (cu yd, yd^3) = 0.76 cubic meter (m^3)

TEMPERATURE (EXACT)
$[(x-32)(5/9)]\ °F = y\ °C$

METRIC TO ENGLISH

LENGTH (APPROXIMATE)
- 1 millimeter (mm) = 0.04 inch (in)
- 1 centimeter (cm) = 0.4 inch (in)
- 1 meter (m) = 3.3 feet (ft)
- 1 meter (m) = 1.1 yards (yd)
- 1 kilometer (km) = 0.6 mile (mi)

AREA (APPROXIMATE)
- 1 square centimeter (cm^2) = 0.16 square inch (sq in, in^2)
- 1 square meter (m^2) = 1.2 square yards (sq yd, yd^2)
- 1 square kilometer (km^2) = 0.4 square mile (sq mi, mi^2)
- 10,000 square meters (m^2) = 1 hectare (ha) = 2.5 acres

MASS - WEIGHT (APPROXIMATE)
- 1 gram (gm) = 0.036 ounce (oz)
- 1 kilogram (kg) = 2.2 pounds (lb)
- 1 tonne (t) = 1,000 kilograms (kg) = 1.1 short tons

VOLUME (APPROXIMATE)
- 1 milliliter (ml) = 0.03 fluid ounce (fl oz)
- 1 liter (l) = 2.1 pints (pt)
- 1 liter (l) = 1.06 quarts (qt)
- 1 liter (l) = 0.26 gallon (gal)
- 1 cubic meter (m^3) = 36 cubic feet (cu ft, ft^3)
- 1 cubic meter (m^3) = 1.3 cubic yards (cu yd, yd^3)

TEMPERATURE (EXACT)
$[(9/5)y + 32]\ °C = x\ °F$

QUICK INCH - CENTIMETER LENGTH CONVERSION

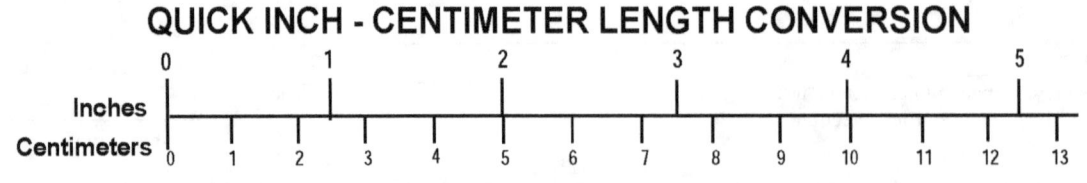

QUICK FAHRENHEIT - CELSIUS TEMPERATURE CONVERSION

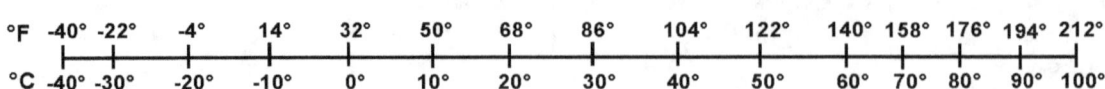

For more exact and or other conversion factors, see NIST Miscellaneous Publication 286, Units of Weights and Measures. Price $2.50 SD Catalog No. C13 10286

ACKNOWLEDGEMENTS

This report was prepared by the Behavioral Safety Research and Demonstration Division of the Human Factors Research and Systems Applications Center of Innovation at the John A. Volpe National Transportation Systems Center (Volpe Center). It was completed with funding from the Federal Aviation Administration's (FAA) Human Factors Research and Engineering Group (AJP-61) in support of the Aircraft Certification Service Avionic Systems Branch (AIR-130) and the Technical Programs and Continued Airworthiness Branch (AIR-120). Our FAA program manager is Dr. Tom McCloy, and our technical sponsors are Colleen Donovan and Bill Kaliardos. We would like to thank Jim Duke, David Gray, Peter Moertl, Don Walker, and Pat Zelechoski for their feedback on this report. We would also like to thank ACSS, Honeywell, and US Airways for allowing us to attend and observe the SURF IA demonstrations and to Blake Kelly at Embry-Riddle Aeronautical University for demonstrating ERAU's surface moving map and ADS-B technologies.

The views expressed herein are those of the authors and do not necessarily reflect the views of the John A. Volpe National Transportation Systems Center, the Research and Innovative Technology Administration, or the United States Department of Transportation.

Feedback on this document may be sent to Michelle Yeh (Michelle.Yeh@faa.gov).

TABLE OF CONTENTS

Executive Summary v
1 Introduction 1
2 Information Sources 3
    2.1 ADS-B Surface Conflict Detection Algorithm Demonstrations 3
    2.2 Use of ADS-B on Airport Surface Moving Maps in a General Aviation Environment ... 3
3 Results 6
    3.1 Use of color 6
        3.1.1 Background 6
        3.1.2 Human Factors Concerns 6
        3.1.3 Recommendations 7
        3.1.4 Research Topics 7
    3.2 Indications/Alerts 8
        3.2.1 Background 8
        3.2.2 Human Factors Concerns 8
        3.2.3 Recommendations 9
        3.2.4 Research Topics 9
    3.3 Traffic Symbols 9
        3.3.1 Background 9
        3.3.2 Human Factors Concerns 10
        3.3.3 Recommendations 11
        3.3.4 Research Topics 11
    3.4 Information Overlay 12
        3.4.1 Background 12
        3.4.2 Human Factors Concerns 12
        3.4.3 Recommendations 12
        3.4.4 Research Topics 13
    3.5 Position Accuracy 14
        3.5.1 Background 14
        3.5.2 Human Factors Concerns 14
        3.5.3 Recommendations 16
        3.5.4 Research Topics 16
    3.6 Air-ground integration 16
        3.6.1 Background 16
        3.6.2 Research Topics 17
    3.7 Other 17
        3.7.1 Background 17
        3.7.2 Human Factors Concerns 17
        3.7.3 Recommendations 18
References 19
Appendix A. Post-Demonstration Questionnaire 20
Appendix B. Surface Moving Map and Traffic Display Usability Questionnaire 22

EXECUTIVE SUMMARY

The purpose of this document is to provide human factors considerations in the integration of traffic information and indications and alerts for runway status on an airport surface moving map. The information is primarily intended to support the development of Federal Aviation Administration (FAA) policy and guidance for surface conflict detection and alerting (e.g., minimum operational performance standards). The US DOT Volpe Center, in support of the FAA Office of Aircraft Certification, gathered information for this document in two ways. First, we conducted observations during demonstrations of ADS-B surface conflict detection algorithms sponsored by the FAA Surveillance & Broadcast Services Office. The intent of the demonstrations was to evaluate the technical feasibility of surface conflict detection and to examine the human factors and safety impacts. However, we were provided with opportunities to observe the displays used during the demonstrations and to talk to the pilots using the displays to gather human factors considerations related to the surface moving map and traffic display. Second, we visited Embry-Riddle Aeronautical University (ERAU) in Daytona Beach, Florida, to understand the impact of a surface moving map with ADS-B in the general aviation operating environment.

The results of this effort provide a preliminary glimpse into potential human factors concerns with the use of a surface moving map, traffic function, and the presentation of surface indications and alerts. The findings are organized into seven categories: use of color, indications/alerts, symbols, information prioritization, airport database, air-ground integration, and other. For each category, a description of the human factors concern, examples, and recommendations for addressing the concern are provided. Additional research will be needed to empirically determine the appropriate design solutions.

1 INTRODUCTION

In 2008, the Federal Aviation Administration (FAA) Surveillance and Broadcast Services (SBS) Office sponsored a program to demonstrate surface conflict detection algorithms and alerting using Automatic Dependent Surveillance – Broadcast (ADS-B) technology. Two contracts were awarded to two teams. One contract was awarded to ACSS and US Airways, who proposed to conduct a demonstration at Philadelphia International Airport (PHL) as well as to equip 20 US Airways Airbus A330s with ADS-B. The second award was to Honeywell, who proposed demonstrations at Seattle-Tacoma Airport (SEA) and Paine Field Airport. The intent of the program was to test and validate requirements for an ADS-B surface application and support the development of minimum operational performance standards.

A surface conflict detection function has two components: (1) an airport surface moving map, which can vary in the level of detail with which the airport surface is depicted, and (2) a traffic function, which can be supported by ADS-B or other surveillance technologies (e.g., Traffic Information System – Broadcast (TIS-B)). Figure 1 presents an example of an airport surface moving map that shows traffic information. In the figure, ownship is represented via a magenta triangle (in the center of the figure), and traffic aircraft are represented via the chevron, diamond, and bullet-shaped symbols.

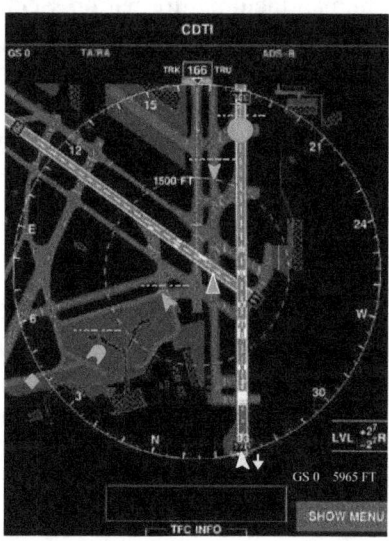

Figure 1. Photo courtesy of ACSS (Excerpted from Yeh and Eon, 2009).

The FAA currently provides guidance for the design and approval of the surface moving map application in Technical Standard Order (TSO)-C165, *Electronic Map Display Equipment for Graphical Depiction of Aircraft Position*, issued on September 30, 2003 and in Advisory Circular (AC) 20-159, *Obtaining Design and Production Approval of Airport Moving Map Display Applications Intended for Electronic Flight Bag Systems*, which was issued on April 30, 2007. TSO-C165 and AC 20-159 reference RTCA DO-257A, *Minimum Operational Performance Standards for the Depiction of Navigational Information on Electronic Maps*, which defines minimum performance standards. RTCA DO-257A applies to equipment that is intended to provide ownship position on an electronic map display, whether it is on the airport surface, in-flight, or vertical situation display. If the surface moving map is presented on an Electronic Flight Bag (EFB), FAA Advisory Circular (AC) 120-76A, *Guidelines for the Certification, Airworthiness, and Operational Approval of Electronic Flight Bag Computing Devices*, and *Human Factors Considerations in the Design of Electronic Flight Bags (EFBs)* by Chandra, Yeh, Riley, and Mangold (2003) may also apply.

The FAA provides guidance for a traffic function in TSO-C195, *Avionics Supporting Automatic Dependent Surveillance – Broadcast (ADS-B) Aircraft Surveillance Applications (ASA)*. TSO-C195 references RTCA DO-317, *Minimum Operational Performance Standards (MOPS) for Aircraft*

Surveillance Applications System (ASAS), which addresses requirements for the presentation of traffic information. The performance standards specified in RTCA DO-257A and RTCA DO-317 define the *minimum* capabilities required for an airport moving map and traffic display, respectively, and higher performance standards may be needed to support more advanced capabilities. Industry input for the development of traffic and runway status indications and alerts is being addressed by RTCA Special Committee (SC)-186 and are documented in RTCA DO-323, *Safety, Performance and Interoperability Requirements Document for Enhanced Traffic Situational Awareness on the Airport Surface with Indications and Alerts (SURF IA)*. RTCA SC-186 is also tasked with developing the MOPS for the surface conflict detection and alerting function.

The FAA Aircraft Certification Service, Avionics Branch has requested human factors guidance to support the development of MOPS for the surface conflict detection and alerting function. Although existing FAA guidance addresses many human factors considerations associated with individual traffic and airport surface map depictions, guidance is needed for the integration of this information. Information for this document was gathered from observations conducted as part of the ADS-B surface conflict detection algorithm demonstrations and information gathered on the use of ADS-B in the general aviation environment. No formal human factors evaluation was planned as part of these ongoing efforts; however, opportunities were provided to gather human factors information. Section 2 describes the sources of information for this report in more detail. The preliminary findings are presented in Section 3. The topics are organized into seven categories: use of color, indications/alerts, symbols, information prioritization, airport database, air-ground integration, and other. Each category contains a description of the human factors issue, examples of the issue, and recommendations for addressing the issue.

2 INFORMATION SOURCES

The Volpe Center gathered information for this preliminary report in two ways:

(1) Observing demonstrations of ADS-B surface conflict detection algorithms sponsored by the FAA Surveillance & Broadcast Services Office.

(2) Observing demonstrations of the airport moving map and traffic display at Embry-Riddle Aeronautical University (ERAU) in Daytona Beach, Florida, and interviewing their flight instructors to understand the impact of these technologies in a general aviation operating environment.

Each of these information sources is described in detail below. The observer notes from the demonstrations and the findings from the discussions with pilots using the displays were summarized and analyzed to identify potential human factors topic areas and the persistence of the usability concerns. The results of this analysis are presented Section 3.

2.1 ADS-B Surface Conflict Detection Algorithm Demonstrations

The Volpe Center observed four demonstrations of ADS-B surface conflict detection algorithms. The first three of these observations were during demonstrations conducted at Philadelphia International Airport (PHL) by ACSS in partnership with US Airways (November 8, December 5, and December 6, 2009). Two aircraft were used in the demonstrations: a US Airways Airbus A330 and an ACSS–owned Beechcraft King Air. The ACSS prototype software was presented on two Electronic Flight Bags (EFBs) installed under the left and right side windows of the US Airways A330 aircraft and on an EFB located between the captain and first officer in the King Air aircraft. Potential conflict scenarios involving the A330 and King Air were conducted in night-time and day-time conditions. The fourth observation occurred during a demonstration by Honeywell at Seattle-Tacoma airport and Paine Field airport in January 2010. Two aircraft participated in the demonstrations: a Cessna Citation Sovereign and a Beechcraft King Air. In the Citation Sovereign, the surface moving map display and proof-of-concept software was presented as part of an integrated navigation display located in front of the captain. On the King Air, the surface moving map and proof-of-concept software was presented on a temporary display, installed specifically for the purposes of the demonstration, and located in front of the first officer.

The primary purpose of these demonstrations was to evaluate the technical feasibility of surface conflict detection and to examine the human factors and safety impacts. During the demonstrations, the Volpe Center took the opportunity to observe the surface moving map and traffic display. During the ACSS-US Airways demonstration, an opportunity was also provided to interview the A330 pilots to gather their opinions on the EFB, surface moving map, and the display of traffic, indications, and alerts. Three of the ACSS-US Airways demonstration participants completed a short questionnaire on the usability of the displays (see Appendix A). Honeywell also conducted a human-in-the-loop simulator evaluation and flight tests of their display concepts in conjunction with their demonstrations; their results are documented in Khatwa and Lancaster, 2010a, 2010b.

2.2 Use of ADS-B on Airport Surface Moving Maps in a General Aviation Environment

To understand the use of a surface moving map and traffic display in a general aviation operating environment, the Volpe Center conducted observations at Embry-Riddle Aeronautical University (ERAU) in Daytona Beach, Florida from March 29-30, 2010. ERAU has 64 aircraft, which are equipped as follows:

Aircraft	Number of Aircraft	Surface Moving Map
Cessna 172 Nav II	2	Garmin MX20
Cessna 172 Nav III	41	Garmin G1000
DA42-L360	10	Garmin G1000
PA44-180	3	Garmin MX20
PA28R	7	Garmin MX20
Decathlon	1	--

Table 1. ERAU aircraft and equippage.

All aircraft both transmit and receive ADS-B information. Neither the Garmin MX20 nor Garmin G1000 provided the capability for airport surface indications or alerts.

The Garmin MX20 and Garmin G1000 have different capabilities with respect to depicting the airport surface. The Garmin G1000 provides the SafeTaxi application, which can show a detailed depiction of the airport surface, if the airport database is available and installed on the aircraft. Some of ERAU's aircraft did not have the full airport database, and in these cases, the surface moving map showed runways only (with centerline markings and runway labels). The surface depiction on the Garmin MX20 consists of runways only.

Information was gathered at ERAU in three ways. First, ERAU provided a demonstration of the Garmin G1000 and Garmin MX20 on three aircraft equipped with the displays – two of the aircraft had the Garmin G1000, the third the MX20. All three aircraft were equipped with ADS-B capabilities. Second, interviews were also conducted with eight of ERAU's instructor pilots who use the displays. The interview sessions consisted of two parts: a paper questionnaire, in which the pilots rated their opinions on the usability of the surface moving map and ADS-B information on the ground (the questionnaire administered is included as Appendix B), and a discussion of human factors concerns related to the use of the surface moving map and/or ADS-B. The questionnaire consisted of Likert-scale items, in which pilots indicated their level of agreement on a scale of 1 (strongly disagree) to 5 (strongly agree). Pilots who completed both an interview and a questionnaire received a $30 gift card for their participation. Note that although eight pilots participated in the interviews, one of the pilots' feedback was excluded because his primary aircraft was the Decathlon, which was not equipped with a surface moving map.

Because some pilots were interested in providing input but could not participate in the interviews, ERAU coordinated the distribution of the paper questionnaire to all of their instructor pilots. The questionnaires were distributed in April, and pilots were given two weeks to complete it. The Volpe Center sent 150 questionnaires for distribution; of these, 44 were returned. (Note that the eight pilots who participated in the interviews were excluded.) Pilots who completed only the paper questionnaire received a $10 gift card for their participation.

In total, 51 Embry Riddle Aeronautical University (ERAU) flight instructor pilots provided input to this effort. The pilots' flight experience is as follows:

Total Hours flown	
1,500 hours or less	24 pilots
1,501 to 3,000 hours	18 pilots
3,001 to 7,000 hours	6 pilots
More than 7,000 hours	3 pilots

Table 2. ERAU Flight Instructor Pilots: Total hours flown.

Most of the pilots (48) primarily used the Garmin G-1000. The other three indicated they primarily used the Garmin MX-20. There is no easy way to judge expertise with a surface moving map. We chose to quantify it by asking participants to estimate the number of taxi segments with which they had operated with a surface moving map. (Note that in this case, taxiing out on departure and taxiing in on arrival are considered two taxi segments.) The majority of pilots (26) had used a surface moving map for more than 200 taxi segments. However, there were 13 pilots who had indicated that they the surface moving map for 50 taxi segments or less.

Estimated number of taxi segments operated with a surface moving map	
10 or less	5 pilots
11 to 50	8 pilots
51 to 100	7 pilots
101 to 200	5 pilots
201 or more	26 pilots

Table 3. Estimated number of taxi segments operated with a surface moving map.

3 RESULTS

The use of a surface moving map showing traffic aircraft, indications, and alerts was generally perceived as being a positive enhancement to the safety of surface operations. In the general aviation environment, all the participants interviewed felt that a surface moving map with ownship supported their position awareness on the airport surface better than a paper chart alone. Several participants attributed this to the depiction of ownship position, which made it easier for them to determine their position on the airport surface. The presentation of traffic information (via ADS-B or TIS-B) was cited as an additional benefit; participants indicated that they generally considered the traffic display to improve their perception of the safety of surface operations.

However, some potential human factors concerns were identified through observations of the different implementations of the surface moving map and traffic displays and through interviews with pilots using the displays. The considerations are organized here into seven topic areas:

   3.1   Use of color
   3.2   Indications/Alerts
   3.3   Traffic Symbols
   3.4   Information Overlay
   3.5   Position Accuracy
   3.6   Air-ground integration
   3.7   Other

Each topic can contain up to four sections: *Background* provides a description of the issue; *Human Factors Concerns* provide examples of the issue; *Recommendations* contain information from FAA regulatory and guidance material or general human factors research regarding how the problem may be addressed; and *Research Topics* list areas where additional research is needed.

3.1   Use of color

3.1.1   Background

Color can be used successfully to annunciate and/or call attention to information on the display. However, the effectiveness of color can be compromised, if implemented inappropriately. In some cases, color may attract attention when none is warranted. The ability to detect a color or to discriminate between colors may be limited not only by one's visual capabilities, but also by external factors, such as the display quality, the display location (e.g., viewing angle, and display position), and the lighting conditions.

The consistent use of color within the surface moving map application and traffic display is highly desirable. There are also a number of color conventions for the use of color for aircraft displays to consider, in particular the use of red and yellow/amber to annunciate warnings and cautions, respectively (see 14 CFR §§ 23.1322, 25.1322, 27.1322, 29.1322).

3.1.2   Human Factors Concerns

- **Ecological color coding.** In many cases, the information elements on the airport surface moving map are colored to resemble the actual object out-the-window. One common application of this color coding is to designate taxiway identifiers in yellow because they are indicated out the window as black text on a yellow border. (This implementation, observed on one of the surface moving maps during the demonstrations, is generally intended to provide a cognitive benefit by improving the pilot's ability to correlate taxiway symbology on the surface moving map display to taxiway signage out-the-window.) This use of yellow/amber is inconsistent with FAA policy and guidance material that recommends those colors be reserved for items indicating the possible need for future corrective actions (see 14 CFR §§ 23.1322, 25.1322, 27.1322, 29.1322).

- **Distracting color changes.** Color may be used on traffic symbols to indicate whether an aircraft is in the air or on the ground. Often, aircraft on the ground are colored as tan/brown, whereas aircraft in the air are cyan. The change in the color of the traffic symbol when an aircraft transitions from the air to the ground (or vice versa) is salient. In fact, during one of the demonstrations, this color change indicating the change in aircraft state was observed to be more attention-getting than the onset of a traffic indication. Thus, this color change may further attract attention, even when it is not intended to do so. It will be important to understand the implications of this color change, particularly during routine operations when aircraft are repeatedly taking off and landing.

- **Use of blue.** Some colors may not be appropriate for indications/alerts. For example, the use of pure blue was used to highlight the runway to indicate a potential runway or traffic conflict in the demonstration displays. However, the presentation of pure blue on a black background (e.g., when outlining a runway) may not be easy to see (i.e., salient) at all map ranges. In general, the use of blue on avionics displays may be problematic. From a physiological standpoint, blue is the shortest wavelength, so it is difficult to bring blue display elements into focus when it is used in combination with other colors. Additionally, as the eye ages, the perception of "blue" may also change, because the lens yellows and absorbs more short wavelength light, affecting the amount of blue light that reaches the retina of the eye. Testing should be conducted to ensure that the color chosen is noticeable to all pilots when viewed briefly.

- **Afterimages.** In the implementations of runway status indications/alerts presented during the demonstrations, a border is drawn around the active runway to highlight it, e.g., with a blue, yellow, or red border. These borders can lead to an afterimage; for example, if the active runway is outlined with a blue border, a yellow afterimage – the illusion of a yellow border surrounding the runway – may result when the blue border disappears. Afterimages are caused when photoreceptors in the eye (in this case, the cones) adapt to a specific stimulus (e.g., the color blue) and lose their sensitivity. Once the color to which the cones were adapted is removed, the location where the color is presented is replaced by its complementary color – for blue, the complementary color is yellow. Afterimages were observed during the demonstrations even upon glancing at the display. One potential consequence of yellow afterimages is that pilots could become desensitized to the appearance of yellow, and this could result in a delayed response to any caution alerts.

3.1.3 Recommendations

- The use of color should be consistent with FAA policy and guidance material; in particular, the use of the colors amber/yellow and red should be specifically examined. The colors red and amber/yellow are normally reserved for alerting functions. The use of these colors for functions other than crew alerting must be limited and must not adversely affect crew alerting. (See also the Aviation Rulemaking Advisory Committee (ARAC) recommendations for existing 14 CFR § 25.1322, http://www.faa.gov/regulations_policies/rulemaking/committees/arac/media/tae/TAE_ASH_T4.pdf.)

- Pure blue should not be used for small symbols, text, fine lines, or as a background color. [see also Cardosi and Hannon, 1999; Chandra, et al. (2003), 2.4.3]

3.1.4 Research Topics

- Research is needed to identify best practices for the use of color when integrating traffic on surface moving maps. A traffic symbol that will overlay different information elements on an airport moving map may be easier to see on one background color (e.g., runways which may be colored black) than on another (e.g., taxiways which may be colored gray). Recommendations are needed specifying the appropriate foreground-background color/brightness contrasts for viewability. Additionally, the use of color to indicate whether the aircraft is in the air or on the ground should be examined to determine if this color change can be implemented so that it does not attract attention inappropriately, particularly during routine operations when aircraft are repeatedly taking off and landing.

3.2 Indications/Alerts

3.2.1 Background

RTCA DO-323 defines two components to facilitate airport surface conflict detection: *indications* for "a normal operational condition that could become a runway safety hazard" (p. 6) and *alerts* for "non-normal operational situations where collision hazard exists or a collision appears imminent" (p. 2).

The RTCA SC-186 WG-1 concept for implementation of the indications and alerts consists of highlighting traffic aircraft or runway, developing symbols for information that is offscale or outside of the current display range, presenting text information, and for alerts only, an auditory message. Detailed guidance for alerts can be found in the Recommended 14 CFR 25.1322 and associated Advisory Circular, AC 25.1322-1, *Flightcrew Alerting*, issued on December 13, 2010.

3.2.2 Human Factors Concerns

- **Salience.** Visual indications used to highlight a runway or traffic aircraft on the surface moving map during the demonstrations were not always salient. The indication of a runway conflict was implemented by outlining the runway with a colored border. On one display, the line thickness used for the border was fixed and therefore independent of map range. The colored border appeared thick at high map ranges (i.e., when the map is zoomed out) but appeared thin at low map ranges (i.e., when the map is zoomed in) and was therefore not as salient.

- **Location.** The saliency of a visual indication or alert will be affected by its location on the flight deck, its visibility relative to other visual/auditory displays, and the flight crew's workload. Some surface moving map displays will be shown on a side display (e.g., on an Electronic Flight Bag (EFB)), as was the case during the demonstrations, so any indications/alerts that are presented directly on the display will not appear within the pilot's primary field of view, an area approximately ±15° horizontal and ±15° vertical in front of the pilot in which information is most easily detected. Discussion regarding the location of surface conflict alerting has included integrating their presentation to the aircraft's master caution and warning systems. There is general consensus, however, that the master caution and warning systems are reserved for aircraft-specific failures; consequently, the presentation of a traffic or runway incursion alert in the master caution and warning panel would be inconsistent with current flight deck philosophy. Additionally, master caution and warning systems notify the pilots to malfunctions of avionics displays that are presented within the primary field of view, so integrating indications and alerts pertaining to a display that is outside the primary field of view into the master caution and warning systems could lead to confusion regarding where to look to diagnose the alert.

- **Auditory Alert.** During the demonstrations, an auditory alert was used to direct the attention of the flight crew to visual alerts (i.e., warnings and cautions). However, the volume of the alert must be carefully set. One common problem is the distraction caused by an auditory alert that is too loud. If the volume of the alert is too high, the alert can interrupt the flight crew, create a startle response, mask communications with air traffic control (ATC), or disrupt the flight crew from completing other necessary tasks. Auditory alerts were not used to call attention to visual indications, however; in the future, it will be important to understand the operational effectiveness of visual indications only.

- **Consistency in implementing surface conflict indications and alerts.** Each of the manufacturers participating in the surface conflict detection demonstrations developed their own algorithm defining the operating conditions for presenting an indication or alert. Consequently, minor differences in the implementations of surface conflict indications and alerts were observed across the demonstrations. Inconsistencies in output behavior of surface conflict detection algorithms will make it difficult for pilots to understand the operating conditions under which indications and alerts may be presented and reduce the usability of that information. Recommendations for the general output behavior of the surface conflict detection algorithms are provided in the *Safety Performance and Interoperability*

Requirements Document for Enhanced Traffic Situational Awareness on the Airport Surface with Indications and Alerts (SURF IA).

- **Training.** The complexity of the presentation of runway indications and alerts will require manufacturers and operators to consider how to optimize the training for this function. A surface conflict detection algorithm may have several states, including normal operations, indications, cautions, and warnings. Each of these states is represented differently, and there is the potential to confuse these different states. Pilots must understand the symbology, the meaning of the attributes used, and the rules in which the runway indications and alerts are presented. As the complexity of any of these factors increases, more extensive training will be needed. Additionally, training to inform pilots about the design philosophy and the thresholds used to identify the circumstances under which indications and alerts occur may help the flight crew understand the occurrence of false indications or alerts.

3.2.3 Recommendations

- Alerts and indications should be consistently located in a specific area of the electronic display. Alerts that may require immediate flightcrew awareness should be located in the flightcrew's primary field of view. [AC 25-11A, 31.f.(2); AC 23.1311-1B, 18.1]
- The implementation of visual indications and alerts must be salient at all map ranges.
- The auditory signal should be at least 20 dB greater than the ambient noise level so that it is detected. The auditory signal should be at least 60 dB but less than 135 dB. (McAnulty, 1995-74)

3.2.4 Research Topics

- Research is needed to understand the usability and effectiveness of surface conflict indications and alerts depending on the location in which they are presented on the flight deck. The presentation of alerts on a side display, such as an EFB, may not be sufficient to attract attention during a non-normal operational condition. The inclusion of an aural alert or the presentation of a separate visual alert in the primary field of view may be needed.
- It will also be important to understand pilot's opinions of the utility and usability of the indications and alerts, particularly in normal operations when there are a large number of other aircraft on the airport surface and traffic and runway status indications may appear relatively frequently. The demonstrations were scheduled at times when there would not be many operations of other aircraft to minimize the impact of the demonstrations on the airports' operations. As a result, no observations were made as to the utility of indications in calling attention to the presence of potential conflict aircraft nor to the frequency of false/nuisance traffic indications or runway status indications in a normal operating environment. Indications or alerts that are perceived to be a nuisance can increase the flight crew's workload and reduce the flight crew's trust. This distrust may cause the flight crew to respond slower to true alerts (i.e., when a non-normal operational or airplane system condition exists), to take an inappropriate action, or fail to act when it is necessary.

3.3 Traffic Symbols

3.3.1 Background

There are several properties for traffic aircraft that may be depicted, such as whether the aircraft is in the air or on the ground, whether the aircraft is in close proximity to ownship, the aircraft's directionality, and the aircraft's reliability. However, there are a limited number of symbol attributes which can be used to convey this information, such as shape, color, or fill. When designing new methods for presenting symbology, it is important to consider consistency with applicable standards, as well as related standards, such as standards for TCAS (Traffic Alert and Collision Avoidance System). Additionally, it will be important to ensure that the symbol set is easily understood by pilots (unambiguous), and that all symbols are distinctive and legible.

3.3.2   Human Factors Concerns

- **Similarity**
    - **In color/intensity**. Similarity of one symbol to other information elements, e.g., in terms of color or intensity, can reduce the saliency of the symbol. In particular, there tends to be an abundant use of white or shades of grey on surface moving maps. Two examples of similarity reducing the saliency of one or more information elements were observed during the demonstrations. The first concerned the depiction of ownship symbol and other aircraft. A common color for depicting ownship symbol is white, but when other information elements on the surface moving map are also depicted in white or other light colors (e.g., gray, light gray, or other light shades), the saliency of the ownship symbol is reduced. Similarly, if traffic aircraft is depicted using white or gray chevrons, it will be important to ensure that these symbols are easy to see against other information elements of similar shades. Second, if the color or intensity used to depict runways is similar to that for taxiways, then the depiction of runways may not be distinctive. In general, considerations regarding the use of color or intensity for individual presentations of traffic symbology and airport moving map information elements may be different than when these information elements are integrated.
    - **Of information elements.** Symbols representing traffic aircraft should be easily distinguishable from other symbology (e.g., navigation aids) or information elements. The surface moving map function may be integrated with a navigation display that shows navigation aid symbology when ownship is in the air. On one display, symbols for traffic aircraft and navigation aids were drawn with similar attributes, e.g., similar line thicknesses, size, and background color (both black). Additionally, on that display, the chevron symbol, used to represent a traffic aircraft, was similar in size and shape to the letter "A" (taxiway identifier). Differentiating symbols along at least one salient attribute can help reduce the amount of information that needs to be searched and reduce the time it takes to find a symbol of interest. (See also Section 3.3.4 for a discussion of related research considerations addressing the design of symbology for the traffic function.)
- **Consistency across avionics displays.** The properties of aircraft that are depicted and the method for depiction may differ from one avionics display to another, even when the avionics displays are developed by one manufacturer. For example, color may be used by two different avionics displays to indicate whether an aircraft is in the air (cyan) or on the ground (tan/brown), but one avionics display may use this coding scheme only for traffic aircraft that are within a certain distance from ownship whereas another uses it for all aircraft shown on the display. Another attribute which has been used inconsistently across avionics systems is fill: one avionics display uses fill to indicate the aircraft that is the selected target whereas another uses it to indicate those aircraft that are in close proximity to ownship. The amount of information provided in a data block describing traffic aircraft also differed from one avionics display to another. Consequently, pilots who fly different types of airplanes use different avionics systems and may not know what information is readily available. Recommendations for conveying different traffic symbol attributes are provided in RTCA DO-317. Inconsistency in the properties of a symbol that are conveyed and how they are conveyed can increase the likelihood of confusing one symbol for another and increase the potential for error. Additionally, unless pilots understand the rules completely, it may not be intuitively clear why a traffic symbol appears one way versus another.
- **Symbol size.** The legibility of the symbol must preserve key features or attributes of the symbol. A symbol's appearance on a display varies depending on the display technology, such as display contrast, resolution, and size. Symbol size will be constrained by the display size, but fine details may be easier to detect and discriminate in large symbols than small symbols. For example, other

manufacturers have proposed embedding a dot into the traffic aircraft symbol (usually a chevron) to convey an attribute (e.g., traffic selection or a traffic indication). Depending on the size of the symbol though, this attribute may not be immediately salient.

- **Depiction of off-scale traffic.** The position of the selected aircraft with respect to the surface moving map when it is off-scale (i.e., when it falls outside the current display range/zoom level) may be indicated by depicting the relative bearing of the selected aircraft, the projected track of the selected aircraft, or both. Showing the relative bearing of the traffic symbol with respect to ownship is consistent with TCAS conventions when ownship is in the air, but may be inconsistent with the actual track of the selected aircraft with respect to a surface moving map. In other words, when airport information elements are used for reference, the depiction of relative bearing could provide misleading information regarding the selected aircraft's actual position, e.g., a traffic aircraft, on approach to an airport, could be shown approaching on one runway when it will in fact land on a different runway. Depicting projected track instead will provide an accurate representation of where the selected aircraft will be in the future (e.g., during the landing scenario), although depiction of the selected aircraft's current position may not be as precise. An understanding as to the level of precision required for the different tasks on or near the airport surface and how to ensure that the presentation of other aircraft is consistent with current technologies (e.g., TCAS) may facilitate the introduction and use of this application.

3.3.3 Recommendations

- The depiction of runways shall be distinctive from all other symbology. [TSO-C165/RTCA DO-257A, 2.3.1.1.1]

  Note: The use of color as the sole means of distinguishing runways may not be sufficient to meet this requirement.

- If traffic is on the ground, the basic traffic symbol shall be modified by changing the color. [RTCA DO-317, 2.3.4.2.3.2.5] (See also Section 3.3.4 for a discussion of related research considerations for indicating the air/ground transition.)

- The symbols used to depict traffic aircraft and the rules for depiction should be consistent across avionics displays.

- The method for representing the position of selected aircraft with respect to the surface moving map when it is off-scale should be consistent within the flight deck for surface operations.

3.3.4 Research Topics

- Guidance is needed to address the characteristics that should be distinguished in the symbol set within the constraints of the recommendations provided in RTCA DO-317 (e.g., aircraft in air vs. aircraft on the ground, aircraft on the ground vs. other ground vehicle). In particular, it is important to understand the interactions among display elements of different colors and the color contrast between the depiction of traffic aircraft and the airport surface moving map. Related to this issue is how the transition for aircraft in the air to that on the ground is presented. The use of color to indicate this transition (e.g., from cyan in the air to brown on the ground) is likely an effective means to easily distinguish the two states without adding complexity to the symbol set. However, the air/ground transition may be overly salient, and inappropriately attract attention. It will be important to understand the implications of such designs; the use of color may still be an effective approach if it can be implemented in such a way that the color change does not inappropriately attract attention.

- Research is needed to understand the precision with which selected traffic aircraft should be depicted with respect to the surface moving map when it is off-scale. The assessment should compare pilot performance when the off-scale position of the selected aircraft with respect to the surface moving map is represented via the selected aircraft's relative bearing versus its projected track. In particular, pilot performance should be considered to determine if there are any consequences when the depiction of off-scale traffic is inconsistent with current implementations.

3.4 Information Overlay

3.4.1 Background

The airport surface environment is rich in detail, so it is likely that when information elements are drawn on the surface moving map and the presentation of traffic aircraft is included, some information elements will be obscured. If information elements are not easily readable, pilots may not be able to obtain information quickly and accurately; this could potentially lead to errors and increase the amount of time spent head-down in the flight deck.

3.4.2 Human Factors Concerns

- **Clutter.** The saliency of a symbol will depend on the amount of information presented on the display (i.e., global density) as well as the amount of information in close proximity to the target symbol (i.e., local density). One concern with showing traffic aircraft on a surface moving map is the number of aircraft that may be depicted, which could contribute to display clutter. Display clutter was not a concern during the surface conflict algorithm demonstrations, but the number of traffic aircraft on the airport surface at the same time was purposefully limited by conducting the demonstrations at off-peak hours. In the general aviation environment, ERAU pilots indicated that display clutter was generally not a concern when acquiring traffic on a surface moving map, but one pilot specifically noted that the airport surface can appear to be a cluster of traffic aircraft when the surface moving map is viewed at high map ranges (i.e., zoomed out). A filtering scheme to prioritize the traffic aircraft may be useful to prevent the perception of display clutter regardless of the map range.

- **Overlaying information.** Information elements will undoubtedly overlap on the surface moving map display, e.g., a traffic identifier will at times appear in the same location as a runway or taxiway identifier. On one prototype display, text labels were sometimes drawn over each other so that the information was not readable, e.g., a traffic identifier drawn over a runway label, or compass rose labels drawn over the runway labels. One convention used on another display was to prioritize the presentation of text labels, so that if two pieces of text information overlapped each other, one had priority while the other was hidden.

- **Associating aircraft and identifiers.** When there are several aircraft in close proximity, the association between an aircraft and its identifier and/or data tag may not be clear. During the demonstrations, there were several aircraft transmitting ADS-B, but not all the aircraft transmitted their identifier. When two aircraft were in close proximity with only one transmitting its identifier, it was not always clear with which aircraft the identifier was associated. If the association between an aircraft and its identifier is ambiguous, the pilot may spend more time heads-down and the potential for error increases (e.g., if the pilot associates an identifier with an incorrect aircraft).

- **Readability/legibility.** Text or symbols may not be readable or legible if it is presented over information elements that are similar in color. For example, on one display, range information printed in white text was not readable when it overlaid a runway that was depicted using a light gray fill, i.e., a color similar to white.

3.4.3 Recommendations

- To ensure the availability of appropriate information during surface operations, the order of display layer precedence (in case aerodrome features overlap) should be (higher priority layered on top): [TSO C-165/RTCA DO-257A, 2.3.4.1]
    (a) Ownship symbol
    (b) Taxi route

(c) Runway identifiers
(d) Runways
(e) Taxiway identifiers
(f) Taxiways

Authors' Note: Aerodrome map features are listed from high to low priority.

- To ensure visibility of the most important symbols and their associated proximate data (i.e., data tags), the following prioritization should be followed for information overlay, from highest to lowest priority: [RTCA DO-317, 2.3.6.1]

  1. Own-ship
  2. Airborne traffic with warning level alerts (including TCAS RAs)
  3. Surface traffic with warning level alerts (see Note 2).
  4. Airborne traffic with cautions level alerts (including TCAS TAs)
  5. Surface traffic with cautions level alerts (see Note 2).
  6. Airborne coupled traffic.
  7. Surface coupled traffic (see Note 2).
  8. Airborne selected traffic.
  9. Surface selected traffic.
  10. Other airborne traffic
  11. Other surface traffic
  12. Other CDTI features.

  Notes:
  1. Altitude, range, and other information may be used for further prioritization and tie breaking, as appropriate for the application.
  2. The initial Aircraft Surveillance Applications (ASA) applications do not require coupling or alerting on the airport surface.

  Authors' Note: The inclusion of indications and alerts in the list above is not yet established.

- Drawing text labels on an opaque text box that is sized to fit the text and/or developing a prioritization scheme to determine what should be depicted when text or display elements obscure each other may help support readability when a lot of information must be shown.

- Traffic identifiers and tags on a display should not obscure each other. The display of tags should be prioritized according to significance to ownship position and route. [SAE ARP 5898, 9.4.3.12]

- Close proximity between an identifier and the aircraft it references can create a positive association between the two. However, when several aircraft are displayed at once, proximity may not be enough to prevent ambiguity. Color can be used to create a positive association between a selected aircraft and the data block for that aircraft, e.g., by matching the text color for information in a data block with the fill color of the corresponding traffic symbol. Drawing a leader line to connect the two may also help, although adding another information element to the display could increase display clutter.

3.4.4 Research Topics

- Guidance for the presentation of traffic information and how it can be integrated with the surface moving map without increasing the perception of display clutter is needed. This guidance should consider the following:
  - An overlay prioritization scheme for the information elements in the presentation of the airport surface and traffic aircraft.

- A filtering scheme (e.g., show or hide) for the display of traffic aircraft since the presentation of all traffic aircraft on or near the airport surface will most likely result in clutter.
- A filtering scheme for the display of airport surface information elements

3.5 Position Accuracy

3.5.1 Background

The positional depiction of ownship, traffic, and airport information on a surface moving map must be accurate enough for the intended function, such as to support the presentation of traffic and runway status indications and alerts. Several errors can contribute to the accuracy of ownship or traffic depiction on a surface moving map display (e.g., position error, latency, survey error, and display resolution). The accuracy requirements for ownship position on a surface moving map are documented in RTCA DO-257A, *Minimum Operational Performance Standards for the Depiction of Navigational Information on Electronic Maps*. The accuracy requirements for the depiction of traffic on a surface moving map display are described in RTCA DO-317, *Minimum Operational Performance Standards (MOPS) for Aircraft Surveillance Applications System (ASAS)*. More stringent accuracy requirements have been recommended by RTCA SC-186, Working Group (WG)-1 to support the display of indications and alerts (See RTCA DO-323, *Safety, Performance and Interoperability Requirements Document for Enhanced Traffic Situational Awareness on the Airport Surface with Indications and Alerts*). Identifying the potential for error was not the focus of the effort, but the demonstration observations and general aviation pilot interviews provide examples to consider in the implementation of a surface moving map and/or ADS-B.

3.5.2 Human Factors Concerns

- **Ownship Position Errors.** General aviation pilots indicated through the interviews and questionnaires that the information shown by the surface moving map usually matched the view out-the-window. However, 11 pilots indicated that they had noticed position errors of ownship, traffic, or the airport map at two airports: Daytona Beach International Airport (KDAB) and Ormond Beach Municipal Airport (KOMN). The types of error detected and number of reports at these airports is listed in Table 4.

Airport	Error Type	# Errors
Daytona Beach International Airport (KDAB)	Ownship was drawn on or near the edge of my taxiway	1
	Ownship was drawn in the grass	3
	Other	5
Ormond Beach Municipal Airport (KOMN)	Ownship was drawn on or near the edge of my taxiway	1
	Other	1

Table 4. Errors on the Surface Moving Map.

As shown in the table, nine errors were reported at Daytona Beach International Airport (KDAB). Of these, one pilot noticed ownship drawn on or near the edge of the taxiway, although comments provided indicated that the error was slight. Three pilots noticed ownship drawn in the grass. Five pilots noted errors that were classified as "other"; these errors included shadow "ownship" targets, errors in ownship heading, and incorrect depictions of traffic aircraft (e.g., due to time delay). Of significance is that one participant mentioned that a traffic aircraft that was holding short of Runway 25R at KDAB was depicted on the surface moving map as being *on* the runway.

Two errors were noted at Ormond Beach Municipal Airport (KOMN). For one error, ownship was drawn on or near the edge of a taxiway. The pilot reporting this error provided comments indicating that the depiction was in a location so that the aircraft was positioned in-between a taxiway and a runway. The second error was categorized as "other".

During the interviews, the participants noted that these errors were rare. Some tended to occur when the system was first turned on or because the surface moving map database was not up-to-date; for example, several of the participants mentioned that Taxiway P5 is closed but still shown on the surface moving map.

It is important to acknowledge that although no other errors were noted at other airports that the participants tended to fly to/from, in many cases, these small airports were depicted with runways only and not in full detail on the surface moving map. That is, a full database was not always available for airports in the surrounding Daytona Beach area, so information elements such as taxiways, ramp areas, and grassy areas, where database errors may be more common than on runways, were not shown.

- **Consistency with information out-the-window.** The surface moving map may not show all aircraft on the airport surface; some aircraft may not be equipped with a surveillance technology, some do not have their transponders turned on, and others may not be visible with other surveillance technology. Furthermore, technical limitations can affect the completeness of the traffic picture. There were several instances in all the demonstrations in which an ADS-B signal was lost and aircraft which were on the airport surface did not appear on the surface moving map. In one rare case, there was also a "false" aircraft target that appeared on the surface moving map, which did not exist out-the window. (Note that information acquired after the demonstration revealed that this aircraft target did not meet the data quality parameters recommended in RTCA DO-323 for presenting the aircraft target on the surface moving map display. The aircraft target reported erroneous values (zero) for the Navigation Integrity Code (NIC), the Navigation Accuracy Code for Position (NACp), and the Surveillance Integrity Level (SIL). See also Section 3.5 Position Accuracy.)

- **Traffic aircraft position errors.** Inaccuracies in the depiction of traffic position on the airport surface may result in the erroneous depiction of an aircraft on one runway when it is in fact on another or the depiction of an aircraft on a taxiway when it is in fact on a runway. Such errors can have human factors implications for the usability of the traffic display by affecting the level of trust pilots place in the information. A few errors were noted throughout this effort. During the demonstrations, there were a few instances when the ADS-B signal was lost, potentially due to reflection from nearby buildings. There was also one instance where a "false" target was observed on the active runway when ownship was on final approach, because that aircraft target was transmitting values of zero for the NIC, NACp, and SIL. In the general aviation environment, participants noted observing errors in the depiction of traffic as a result of time delay, technology (e.g., when equipped with a Mode A transponder, the use of TIS-B compared to ADS-B), or loss of traffic directionality information.

- **Full airport display vs. runways-only depiction.** ERAU pilots commented on the usefulness of showing the full airport on the surface moving map relative to a depiction of runways only. A runways-only depiction of the airport does not require a full airport database, but there are concerns that showing traffic without a map context can be confusing to pilots regarding traffic positions. Some of the Garmin G1000 Safe Taxi displays and all of the Garmin MX 20 provide only runways on the surface moving map. A runways-only depiction was not an issue for any of these participants, because they fly to/from the same airports and are therefore very familiar with those airports. Additionally, one of the participants noted that most of the airports they fly to/from also have simple layouts, and in many cases, the taxiways are parallel to runways, so it is easy to infer position. Showing traffic information on a runway-only surface moving map display was still considered to be useful, although one pilot noted that such a depiction sometimes required reference to the paper chart.

3.5.3   Recommendations

- Pilots should be trained about system limitations and database errors, such that in the case of any differences between the information on the surface moving map display and the out-the-window view, the information out-the-window takes precedent (14 CFR § 91.113).
- Text in the pilot's guide and airplane flight manual should document the inaccuracies in the presentation of traffic position and which part of the traffic symbol (e.g., apex of arrowhead) corresponds to the aircraft's actual position.
- Pilots should be informed and educated to the display of traffic and presentation of surface conflict detections in a mixed equipage environment. Pilots should be trained to understand at which locations these errors are most likely to occur and provided a process to report any errors to the appropriate authority so that the error can be validated.
- Pilots should be trained on the accuracy of the different surveillance technologies.

3.5.4   Research Topics

- There are many potential sources of error in the depiction of traffic aircraft on a surface moving map display (e.g., surveillance error, positioning error, database accuracy and resolution, survey accuracy and resolution, latency). It will be important to understand the likelihood of these errors and how degraded information should be depicted on the surface moving map display. TSO-C165/RTCA DO-257A provides recommendations for indicating the potential for error in the depiction of ownship position, for example, by using a circle of uncertainty. However, these guidelines were developed for an airport surface moving map with ownship-only display, so the designer should be careful about applying these same conventions when a traffic function is added. That is, some conventions that are appropriate for an ownship-only display may adversely affect the usability of the information when traffic aircraft is depicted (a circle of uncertainty).
- RTCA DO-323 specifies accuracy requirements for the depiction of traffic aircraft, indications, and alerts. Traffic aircraft, indications, and alerts are *not* intended to be presented if aircraft do not meet their respective accuracy and other data quality requirements. Research should examine the value of depicting the accuracy level of the traffic aircraft on the surface moving map and to determine if such depictions are identifiable and useful. In particular, it will be important to determine if pilots can implicitly understand for which aircraft indications and alerts may be presented, which ones do not qualify for indications or alerts, and why.
- Additional research is also needed to determine if a runway-only map is sufficient for the applicable intended functions. The general aviation pilots who provided input into the usefulness of showing traffic information on a full airport depiction versus a runways-only depiction fly to/from the same airports and are thus very familiar with the layouts of those airports. Additionally, the layouts of the airports frequented by the general aviation pilots who participated in this effort were generally such that the taxiways were parallel to the runways so their position on the airport surface was easy to infer. The presentation of aircraft traffic information on a runways-only depiction at airports where an aircraft's position relative to a runway may not be so easily inferred (e.g., if there are several taxiways parallel to a runway) or at an airport that is not familiar to the pilot or flightcrew should be considered. Guidance is also needed regarding the presentation of airport surface indications and alerts on a runways-only display relative to a full airport surface map display.

3.6   Air-ground integration

3.6.1   Background

Some airports also already incorporate ground-based alerting technologies, such as runway status lights, or a hybrid of ground-based and flight-deck system, such as ASDE-X, to provide direct alerts about runway occupancy. If the algorithms used by the ground-based system to determine an alert condition differ from flight-deck based systems, there may be occasions where one system generates an alert and the other does not. Such inconsistencies could create a distraction for the flight crew and increase flight crew workload as pilots diagnose the situation.

Additionally, a surface moving map that shows traffic aircraft and surface conflict indications and alerts may change the nature of pilot communication with air traffic control. General aviation pilots indicated that the presentation of other traffic information positively influenced their discussions with ATC. These pilots noted that taxi instructions are easier to visualize, and the presentation of traffic helped them track – and in one case, find – the aircraft that ATC requested they follow. Several participants also noted that they sometimes clarified instructions with ATC based on the traffic information they had available. In some cases, the surface moving map and traffic display served as a planning tool for pilots. The availability of traffic information helped general aviation pilots identify the congested areas on the airport surface or the busy runways. On final approach, these pilots indicated that the traffic display helped them to see the traffic pattern of other aircraft in relation to their aircraft and sometimes helped them determine where they should be to line up for the approach or motivated them to ask for deviations in their approach pattern.

It needs to be determined what information air traffic control needs regarding the presentation of flight deck alerts. One concern is that the pilot may contact a controller if s/he sees another aircraft, indication, or alert on the surface moving map and question the clearance that was provided.

3.6.2 Research Topics

- It will be important to consider the impact that this technology will have in terms of coordination with air traffic control. Although pilots' questioning of air traffic controller instructions may be routine, the explicit presentation of an alert or indication may increase pilot interactions with air traffic control relative to current operations. The potential impact runway indications and alerts will have on communications on the flight deck and with air traffic control as well as the potential workload imposed on air traffic controllers should be examined. Additionally, it needs to be determined what, if any, information air traffic control needs regarding the presentation of flight deck alerts to prevent any uncoordinated aircraft movement.

3.7 Other

3.7.1 Background

Some general human factors considerations associated with the display readability, the presentation of runway/taxiway identifiers, and the potential impact of a surface moving map with traffic on pilot workload is noted in this section.

3.7.2 Human Factors Concerns

- **Display readability**: The avionics display (and all its colors) should be bright enough so that it is readable during daylight without having the sun shade down. Although drawing the shade can reduce glare and improve readability in bright sunlight conditions, the shade can not be used during take-off and landing.
- **Runway/taxiway identifiers**: In some implementations of surface moving maps, runway and taxiway identifiers are drawn in such a way that their position is fixed relative to the corresponding runway or taxiway. For runways, identifiers are placed at the end of the runway; consequently, when the pilot zooms in to the surface moving map (i.e., reducing the map range), the runway identifiers may not be in view. For taxiways, the fixed location of identifiers on a taxiway may prevent the pilot from knowing which taxiway s/he is approaching depending on the map range, the length of the taxiway, and the location of the identifier relative to the taxiway. If the labels are placed only at runway ends, at runway and/or taxiway intersections, or spaced at pre-defined intervals, the pilot may not be aware

of ownship position relative to runways and taxiways. This could increase heads-down time as well as the potential for errors.

- **Workload**: The use of a surface moving map and traffic display may increase workload because additional tasks will be required (e.g., scanning and attending to the information on the display). This increase in workload may be acceptable given the potential benefits provided by the application. Additional research examining how the surface moving map and traffic display moderates pilot's scanning of other flight deck instruments and out-the-window view may be beneficial to understanding any costs or benefits in the use of these displays.

Preliminary subjective data from general aviation pilots who used a surface moving map and traffic display as part of their regular operations indicated that the pilots perceived that their head-down time decreased slightly when using the surface moving map compared to using a paper airport chart only. No objective data was collected, however. One participant commented that he felt the surface moving map reduced head-down time relative to the paper airport chart, because the surface moving map provided an immediate indication of ownship position on the airport surface, whereas with an airport diagram, he needed to spend more time looking at it to determine ownship position.

All the participants indicated that their primary reference when taxiing on the airport surface is the out-the-window view, and that the surface moving map and traffic display was considered supplementary information.

3.7.3 *Recommendations*

- At reduced map ranges, at least one identifier should be displayed for any taxiway or runway depicted within the selected map range. [TSO C-165/RTCA DO-257A, 2.3.2]
- Procedures should be developed to ensure that the overall workload on the flight deck is acceptable with the addition of the EFB and the surface moving map application.

REFERENCES

1. Federal Aviation Administration, Advisory Circular (AC) 20-159, *Obtaining Design and Production Approval of Airport Moving Map Display Applications Intended for Electronic Flight Bag Systems.* April 30, 2007.
2. Federal Aviation Administration, Advisory Circular (AC) 25.1322-1, *Flightcrew Alerting.* December 13, 2010.
3. Federal Aviation Administration, Advisory Circular (AC) 120-76A, *Guidelines for the Certification, Airworthiness, and Operational Approval of Electronic Flight Bag Computing Devices.* March 17, 2003.
4. Federal Aviation Administration, Advisory Circular (AC) 120-86, *Aircraft Surveillance Systems and Applications.* September 16, 2005.
5. Federal Aviation Administration, Technical Standard Order (TSO)-C165, *Electronic Map Display Equipment for Graphical Depiction of Aircraft Position.* September 30, 2003.
6. Federal Aviation Administration, Technical Standard Order (TSO)-C195, *Avionics Supporting Automatic Dependent Surveillance – Broadcast (ADS-B) Aircraft Surveillance Applications (ASA).* September 27, 2010.
7. Khatwa, R. and Lancaster, J. (2010a). *Flight Simulator Evaluation of a Cockpit Display of Traffic Information (CDTI) With and Without Indications & Alerts (IA).* Performed under contract to the FAA under FAA Agreement Number DTFAWA-09-A-00001.
8. Khatwa, R. and Lancaster, J. (2010b). *Human Factors Flight Test Evaluation of a Cockpit Display of Traffic Information (CDTI) With and Without Indications & Alerts (IA).* Performed under contract to the FAA under FAA Agreement Number DTFAWA-09-A-00001.
9. RTCA DO-257A, *Minimum Operational Performance Standards for the Depiction of Navigational Information on Electronic Maps.* June 25, 2003.
10. RTCA DO-317, *Minimum Operational Performance Standards (MOPS) for Aircraft Surveillance Applications System (ASAS).*
11. RTCA DO-323, *Safety, Performance and Interoperability Requirements Document for Enhanced Traffic Situational Awareness on the Airport Surface with Indications and Alerts.*
12. Chandra, D.C., Yeh, M., Riley, V., and Mangold, S.J. (2003). *Human Factors Considerations in the Design and Evaluation of Electronic Flight Bags (EFBs). Version 2.* DOT-VNTSC-FAA-03-07 and DOT/FAA/AR-03/67. Cambridge, MA: USDOT Volpe Center. Available at www.volpe.dot.gov/opsad/efb.

APPENDIX A. POST-DEMONSTRATION QUESTIONNAIRE

1. Were you aware of any misleading position errors of your own aircraft or traffic during this demonstration?

    _____ Yes   _____ No

    If yes, please explain.

2. How did the use of the following affect your perception of the safety of surface operations?

    Surface map with ownship

Decreased safety		No difference		Increased safety
1	2	3	4	5

    Traffic display

Decreased safety		No difference		Increase safety
1	2	3	4	5

    Indications

Decreased safety		No difference		Increased safety
1	2	3	4	5

    Alerts

Decreased safety		No difference		Increased safety
1	2	3	4	5

3. The difference between traffic indications, runway status indications, and alerts was easy to understand.

Strongly Disagree				Strongly Agree
1	2	3	4	5

    If you disagree (1, 2), please indicate why.

Traffic Display

4. The traffic symbols were easy to distinguish from one another and easily understood.

Strongly Disagree				Strongly Agree
1	2	3	4	5

5. The head-down time required for viewing displayed traffic was acceptable.

Strongly Disagree				Strongly Agree
1	2	3	4	5

6. How did the presentation of traffic affect workload compared to what you did without a traffic display?

Increased	2	Did not change	4	Decreased
1		3		5

If you indicated an increase in workload (1 or 2), please indicate why.

Indications/Alerts

7. The presentation of indications or alerts allowed sufficient time to react to the potential conflicts.

Strongly Disagree	2	3	4	Strongly Agree
1				5

8. The <u>visual</u> alerts (runway and traffic highlighting) were easily noticed and understood.

Strongly Disagree	2	3	4	Strongly Agree
1				5

9. <u>Auditory</u> alert messages were easily understood.

Strongly Disagree	2	3	4	Strongly Agree
1				5

10. The volume of the auditory alerts was:

Too Low	2	Just Right	4	Too Loud
1		3		5

11. The indications and alerts were reliable.

Strongly Disagree	2	3	4	Strongly Agree
1				5

If you disagree (1, 2), please indicate what was presented in error.

APPENDIX B. SURFACE MOVING MAP AND TRAFFIC DISPLAY USABILITY QUESTIONNAIRE

INTRODUCTION

This survey collects your opinion on the surface moving map and traffic display so that we can understand the impact of these technologies on safety. We expect that it will take less than 15 minutes to complete. Most of the questions request a response on a scale from 1 to 5. However, any additional information you would like to provide would be appreciated and useful to help us understand your opinions.

This survey is being conducted by the US DOT Volpe Center, under an agreement with the FAA Human Factors Research and Engineering Group in support of the FAA Office of Aircraft Certification. If you have any questions, please contact:

    Michelle Yeh   Michelle.Yeh@dot.gov 617.494.3459
                          or
    Scott Gabree   Scott.Gabree@dot.gov 617.494.2530

---

1. Were you aware of any position errors of ownship, traffic aircraft, or the airport map?
   _____ Yes   _____ No

   If yes:
   At which airport (4-letter ICAO identifier)? _____

   What was the most significant position error of ownship or the airport map that you observed at that airport?
      _____ No errors.
      _____ Ownship was drawn on the wrong **runway**
      _____ Ownship was drawn on or near the edge of my **runway**
      _____ Ownship was drawn on the wrong **taxiway**
      _____ Ownship was drawn on or near the edge of my **taxiway**
      _____ Ownship was drawn in the grass
      _____ Ownship was drawn in the wrong location in the ramp areas
      _____ Other

   Please provide additional information on the position error (e.g., where was your aircraft or other aircraft, where was it shown on the surface moving map, what was the approximate size of the error).

2. How did the use of the following affect your perception of the safety of surface operations?

   Surface map with ownship

Decreased safety		No difference		Increased safety
1	2	3	4	5

   Traffic display

Decrease safety		No difference		Increase safety
1	2	3	4	5

3. The surface moving map provided sufficient awareness of my position with respect to *runways*.

Strongly Disagree				Strongly Agree
1	2	3	4	5

4. The surface moving map provided sufficient awareness of my position when approaching *runway-taxiway intersections*.

Strongly Disagree				Strongly Agree
1	2	3	4	5

5. The surface moving map helped me determine which taxiway I was on.

Strongly Disagree				Strongly Agree
1	2	3	4	5

6. Runways were easily distinguishable from taxiways and other movement areas.

Strongly Disagree				Strongly Agree
1	2	3	4	5

7. The display symbol for my aircraft was easy to identify.

Strongly Disagree				Strongly Agree
1	2	3	4	5

8. The ownship symbol did not interfere with the legibility of taxiway or runway labels.

Strongly Disagree				Strongly Agree
1	2	3	4	5

9. Ownship symbol heading/directionality was accurate:

Never		Sometimes		Always
1	2	3	4	5

   If inaccurate, the heading/directionality was incorrect when the aircraft was: (Check all that apply)
   _____ stationary
   _____ turning
   _____ otherwise moving.

   Please describe the conditions under which this error occurred.

10. The information shown by the surface moving map sufficiently matched what I saw out the window.

Strongly Disagree				Strongly Agree
1	2	3	4	5

11. The surface moving map showed the information I needed to establish, maintain, and regain position awareness on the airport surface.

Strongly Disagree				Strongly Agree
1	2	3	4	5

12. Did you adjust the map range?
    _____ Yes _____ No

    If yes, please answer the following two questions.

    It was easy to adjust the map range.

Strongly Disagree				Strongly Agree
1	2	3	4	5

    The speed with which the map was redrawn when the map range was adjusted was adequate.

Strongly Disagree				Strongly Agree
1	2	3	4	5

13. Use of the surface moving map _____ my understanding of taxi route clearances when communicating with air traffic control.

Interfered with		Did not change		Improved
1	2	3	4	5

14. How did the use of the surface moving map during airport surface operations affect your heads-down time when compared to a conventional paper/electronic chart?

Increased Heads-Down Time		Did not Change		Decreased Heads-Down Time
1	2	3	4	5

15. Use of the surface moving map during taxi _____ the workload associated with taxi operations compared to using a paper/electronic chart alone.

Increased		Did not change		Decreased
1	2	3	4	5

    If you indicated an increase in workload (1 or 2), please indicate why.

Traffic Display

16. The symbols used to depict aircraft on the ground were easily understood.

Strongly Disagree				Strongly Agree
1	2	3	4	5

17. The traffic out-the-window appeared at the same <u>relative position</u> on the surface moving map display.

Strongly Disagree				Strongly Agree
1	2	3	4	5

18. The position of displayed traffic relative to ownship was easy to interpret.

Strongly Disagree				Strongly Agree
1	2	3	4	5

19. Displaying traffic on the surface moving map improved my ability to visually acquire traffic.

    Strongly Disagree   Strongly Agree
         1   2   3   4   5

20. Display clutter was <u>not</u> a problem when trying to acquire aircraft visually.

    Strongly Disagree   Strongly Agree
         1   2   3   4   5

21. I did not confuse the traffic symbols with any other map feature.

    Strongly Disagree   Strongly Agree
         1   2   3   4   5

22. The head-down time required for viewing displayed traffic was acceptable.

    Strongly Disagree   Strongly Agree
         1   2   3   4   5

23. The traffic display facilitated my communications with air traffic control.

    Strongly Disagree   Strongly Agree
         1   2   3   4   5

24. How did the presentation of traffic affect workload compared to what you did without a traffic display?

    Increased   Did not change   Decreased
        1   2   3   4   5

    If you indicated an increase in workload (1 or 2), please indicate why.

Avionics Display

25. The display brightness adjustment was effective in producing an acceptable range of brightness levels in the lighting conditions I encountered.

    Strongly Disagree   Strongly Agree
         1   2   3   4   5

26. The display was in a location where it was easy to use.

    Strongly Disagree   Strongly Agree
         1   2   3   4   5

27. The size of all buttons/controls were easy to use.

    Strongly Disagree   Strongly Agree
         1   2   3   4   5

28. All buttons/controls are labeled consistently with their function.

    Strongly Disagree   Strongly Agree
    1   2   3   4   5

29. The readability of the text on the display was acceptable from my seating position.

    Strongly Disagree   Strongly Agree
    1   2   3   4   5

30. The information on the display was readable in the lighting conditions I encountered.

    Strongly Disagree   Strongly Agree
    1   2   3   4   5

    If you responded 1 (Strongly Disagree), please indicate under which lighting conditions the information on the display was **not** readable.

    ____ Bright sunlight - Sun coming in the forward window
    ____ Bright sunlight - Sun coming in the side window
    ____ Bright sunlight - other
    ____ Low ambient lighting conditions (dawn, dusk, heavy overcast)
    ____ Night flight/overhead light

31. All of the colors could be interpreted under all lighting conditions I encountered.

    Strongly Disagree   Strongly Agree
    1   2   3   4   5

    If you responded 1 (Strongly Disagree), please indicate under which lighting conditions the colors were **not** readable.

    ____ Bright sunlight - Sun coming in the forward window
    ____ Bright sunlight - Sun coming in the side window
    ____ Bright sunlight - other
    ____ Low ambient lighting conditions (dawn, dusk, heavy overcast)
    ____ Night flight/overhead light

32. System processing never slowed to the point where normal use was impaired.

    Strongly Disagree   Strongly Agree
    1   2   3   4   5

Background/Demographics

33. Total Hours Flown:
    _____ 1,500 hours or less
    _____ 1,501 to 3,000 hours
    _____ 3,001 to 7,000 hours
    _____ More than 7,000 hours

    Last 90 days:
    _____ 25 hours or less
    _____ 26 to 75 hours
    _____ 76 to 125 hours
    _____ 126 to 225 hours
    _____ More than 225 hours

34. Please estimate the approximate number of **taxi segments** that you have operated with a surface moving map. (Keep in mind to count two if you used surface moving map at both ends of a flight):

    _____ 10 or less
    _____ 11 to 50
    _____ 51 to 100
    _____ 101 to 200
    _____ 201 or more

35. What surface moving map display(s) do you use most frequently? _____

    Do you use any other surface moving map display? _____ Yes _____ No

    If yes, please indicate which one. _____

36. The training I received on how to use the surface moving map was adequate.

Strongly Disagree				Strongly Agree
1	2	3	4	5

    What additional training would you like?

37. The training I received on how to use the traffic information was adequate.

Strongly Disagree				Strongly Agree
1	2	3	4	5

    What additional training would you like?

www.ingramcontent.com/pod-product-compliance
Lightning Source LLC
Chambersburg PA
CBHW081809170526
45167CB00008B/3385